SUNFLOWERS

J.F. QUICK

look!

SUNFLOWERS

CREATIVE EDUCATION • CREATIVE PAPERBACKS

Published by Creative Education
and Creative Paperbacks
P.O. Box 227, Mankato, Minnesota 56002
Creative Education and Creative Paperbacks are imprints of
The Creative Company
www.thecreativecompany.us

Design and art direction by Blue Design
Edited by Joe Tischler

Photographs by Alamy (Stanislas Merlin), Dreamstime (Jeffrey Thompson, Zuperpups), Getty (Alex_Chugun, anmbph, bgwalker, cheekylorns, dionisvero, epantha, Fotofreak75, gyro, MilenaKatzer, Olga Guchek, proxyminder, Richard Griffin, valentinarr, WieslawFila), Pexels ((Arina Krasnikova, Diana Light, Irina Zhur, Kaboompics .com, Marek Piwnicki, Petr Ganaj, Robert So), Shutterstock (3DMI), Unsplash (Andrew Shelley, Ira Huz, Kelsey Weinkauf, Lucian Andrei, Mike Erskine, NordWood Themes, Önder Örtel)

Library of Congress Cataloging-in-Publication Data

Names: Quick, J. F., author.
Title: Sunflowers / by J.F. Quick.
Description: Mankato, Minnesota : Creative Education and Creative Paperbacks, [2025] | Series: Look! | Includes bibliographical references and index. | Audience: Ages 8-12 | Audience: Grades 4-6 | Summary: "An engaging story of the life cycle of sunflowers, this middle-grade book features eye-catching photographs, insightfully simple language, and biological facts about this popular crop"-- Provided by publisher.
Identifiers: LCCN 2023050082 (print) | LCCN 2023050083 (ebook) | ISBN 9781640268548 (library binding) | ISBN 9781682774045 (paperback) | ISBN 9798889890409 (ebook)
Subjects: LCSH: Sunflowers--Juvenile literature. Classification: LCC SB299. S9 Q53 2025 (print) | LCC SB299.S9 (ebook) | DDC 583/.983--dc23/eng/20231102
LC record available at https://lccn.loc.gov/2023050082
LC ebook record available at https://lccn.loc.gov/2023050083

Printed in China

TABLE OF CONTENTS

Sunflowers are large, bright-yellow flowers that grow on top of tall stems. They grow around the world in fields and gardens. They usually grow close together in groups.

Sunflower seeds have small, black shells. They can be eaten for snacks. Crushed sunflower seeds make oil used in cooking.

The middle of a sunflower is made of thousands of tiny flowers.

Sunflowers can grow in many places around the world. Scientists even grow them in space.

A sunflower begins when a sunflower seed is planted in the ground. Sunflowers grow best in spring when the soil is warmer.

When spring rain soaks the seed, it starts to grow. **Roots** grow down into the earth. A stem grows up out of the ground. The new plant is called a seedling.

roots: the parts of a plant that grow underground

When the seedling grows up into the air, it's called a shoot. Leaves begin to grow on the shoot. The leaves collect light, water, and air to make food for the plant.

Over a few weeks, the stalk begins to grow tall and strong. Sunflowers often grow as tall as a basketball hoop.

The tallest sunflower ever was over 30 feet (9.14 meters) tall. That's as tall as a telephone pole!

Sunflower shoots are often eaten in salads.

A sunflower puts on a colorful show when a large, round blossom opens. This happens about two months after planting. The blossom has bright yellow petals. In the center, it has a dark, fluffy disk. The blossoms produce yellow, sticky **pollen**.

pollen: tiny yellow grains produced by blossoms that help plants make new plants

Each day, sunflowers turn their blossoms to follow the sun as it moves across the sky.

Bees land on sunflower blossoms. They drink a sweet liquid in the blossoms called nectar. Tiny bits of pollen stick to their fuzzy bodies. The bees carry pollen to other sunflowers.

Sunflower need pollen from other sunflowers to **pollinate** their blossoms. When a blossom is pollinated, it begins to create seeds. The seeds grow all summer.

pollinate: to carry pollen from one flower to another

KINDS OF SUNFLOWERS

MAMMOTH

Its seeds are great for snacks and for feeding the birds. It grows up to 12 feet (3.7 m) tall.

TEDDY BEAR

Its small size makes it good for a garden at home. It grows up to 3 feet (0.9 m) tall.

SUNRICH GOLD

Its medium size and bright colors make it best for flower bouquets. It grows up to 5 feet (1.5 m) tall.

BLACK PEREDOVIC

Its seeds are very oily. Farmers grow them to make sunflower oil. It grows up to 6 feet (1.8 m) tall.

Farmers harvesting must use special tools for the shape and size of sunflowers.

In late summer or early autumn, when the sunflowers' heads become heavy with seeds, it's time for harvest.

These seeds can be enjoyed as snacks or crushed to make oil. Sunflower oil is used in baking and cooking. People like it because it's healthy for the heart.

Sunflowers cannot live in cold weather. With the first **frost** in autumn, they will die. Cold weather brings the end of the life cycle for a sunflower.

Even when a sunflower is gone, its seeds can grow many more sunflowers in the future.

frost: when the dew outside freezes on a chilly morning

Every year, sunflowers go through a **cycle**. A seed is planted in the ground. The seedling grows for several weeks. A sunflower begins to blossom after planting. The plant gets pollinated. Seeds appear and grow all summer. Seeds are harvested in autumn. The sunflower dies when it gets cold. Its seeds can make new sunflowers the next year.

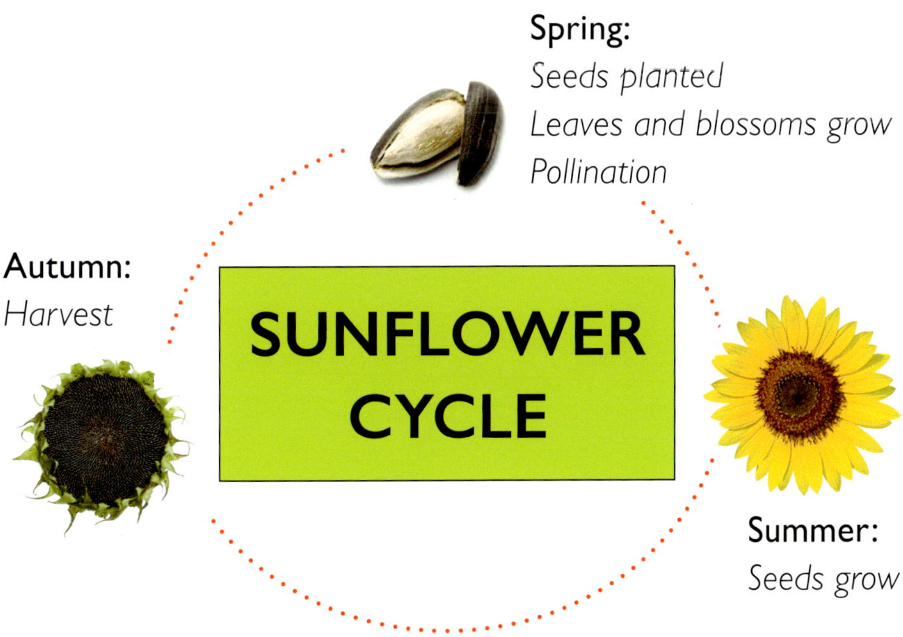

cycle: a pattern that happens over and over in the same order

The sunflowers is the symbol for the country Ukraine. Farmers in Ukraine grow lots of sunflowers. The people of Ukraine are very proud of them. The golden petals of sunflowers look like the sun, which brings warmth and comfort. Sunflowers make the people of Ukraine think of bright, happy days to come.

The world's largest sunflower maze is in England. It is the size of three football fields!

WEBSITES

Grow Sunflowers
https://study.com/academy/lesson/sunflower-lesson-for-kids-facts-life-cycle.html
A simple guide for kids to grow their own sunflowers.

Study Sunflowers
https://www.kidsdogardening.com/how-to-grow-sunflowers/
Become a sunflower expert. Check out a video, lessons, and even a quiz about sunflowers.

READ MORE

Connors, Kathleen. *How Do Sunflowers Grow?*
New York: Gareth Stevens Pub, 2021.

Stewart, Melissa. *A Seed is the Start.*
Washington, D.C.: National Geographic Kids, 2018.

INDEX